Solar Photovoltaic Design for Residential, Commercial and Utility Systems

Contents

1. Introduction

Solar photovoltaic technology has now become mainstream and is widely used around the world. Europe lead the way and now the USA is rapidly catching up. All of the items needed to construct a reliable system can now be purchased from many different vendors and standards now exist in the industry to ensure that these different products can be seamlessly integrated together. The future of solar photovoltaics is bright and rapid adoption of the technology is underway.

This book will demonstrate how to put well built and reliable grid interconnected systems together across the different scales of the applications. As you will see, there really is not that much difference between a small residential system and a large utility scale system.

This book should be read in conjunction with your local building codes, electrical codes and grid interconnection utility codes. These vary widely and will be very much location dependent variables. Always consult with a Professional Engineer who is experienced in solar photovoltaics when building one of these systems.

Once built, you should have many years of reliable operation and very low electricity bills. Solar photovoltaics has longevity with many companies offering twenty five year warranties on their solar modules. It is quite possible that you will only ever purchase one system during your lifetime.

2. The Basics of Solar Photovoltaics

Solar photovoltaics comes in many different types:

- Monocrystalline
- Polycrystalline
- Thin film
- Other technologies

There are many books written on the different types of solar module technologies and we will not duplicate those books here. Although the technology is different between each type, they all do the same thing. All direct current (DC) solar modules generate DC electricity when exposed to sunlight. The listing above is in order of how efficiently each type will convert sunlight into electricity. Solar technology is like a commodity and fluctuates with demand. Generally you would choose the technology that you would use based on the market price at the time of purchase. The less efficient the technology used, the bigger the physical size of the system will be and this will push up the costs of the supporting infrastructure, such as cables, mountings, installation time and so on.

We do not need to understand how the modules work in order to put a system together. Rather, we just need to understand the electrical characteristics of the individual modules to develop the system. A typical direct current (DC) module will have the following electrical ratings on its label:

- Temperature adjustments

- DC Open circuit voltage (Voc)

- DC Maximum power point voltage (Vmpp)

- DC Short circuit current (Isc)

- DC Maximum power point current (Impp)

- DC Rated system voltage

All of these values are given for Standard Test Conditions (STC). Lets look at what each one of these mean:

Standard Test Conditions (STC)

Standard Test Conditions (STC) is how the solar module performs at a temperature of 25 °C, an irradiance of 1000 W/m² with an air mass 1.5 (AM1.5) spectrum. This is a standard test for all solar modules that are manufactured for the USA market that was developed by the photovoltaic industry and the government. It represents an average set of conditions that can be expected at the mid point between North and South of the contiguous forty-eight states during spring and fall. San Francisco, California and Wichita, Kansas are near this midpoint of 37 degrees latitude. In Asia Seoul, Korea, is near and in Europe both Sevilla, Spain and Cantania, Italy are near to 37 degrees latitude.

It is important to note that a solar module output will be continuously variable during the year and even during the day. In wintertime it will output less power than its rating and in summertime it will frequently output more power.

These electrical ratings are for guidance only and it is where many new photovoltaic designers make mistakes in thinking that the module will never output more power than its rating. It is important that you understand that these modules can output far more power than their label states. It can be over fifty percent more and you need to factor this into your design.

Temperature Adjustments

Solar modules are affected by temperature, both hot and cold, and adjustments to the module ratings needs to be made for the operating temperature outside of 25 degrees Celsius. It is important when designing a system that the historical temperature minimum and maximum values are known for the area where the system is to be installed and these adjustments are factored into the design.

Open Circuit Voltage (Voc)

The open circuit voltage rating is how much voltage the module will put out with no load attached. This is an important value in order to design a system. This is the voltage to use when selecting your components and it must be adjusted for the historical minimum and maximum temperatures for the area. If more than one module is connected in series than multiply this temperature adjusted voltage by the number of modules in series to get the total maximum DC voltage of the system.

DC Maximum Power Point

The DC maximum power point is a simple concept. Power is a function of both voltage and current. The maximum power point is obtained when the current and voltage from the module when multiplied together give the maximum power figure. These values will change constantly during the day with the weather conditions. Voltage will remain relatively constant, but current will vary a lot with irradiance.

DC Maximum Power Point Voltage (Vmpp)

The DC maximum power point voltage (Vmpp) is the operating voltage of the solar module under load. Again this value will change with temperature and irradiance, but should only vary by about twenty percent of the STC rating during the day time.

DC Short Circuit Current (Isc)

The DC short circuit current value is the maximum current that the module will output at Standard Test Conditions if the positive and negative terminals were connected (shorted) together. It is important to note that this value will vary a lot dependent on weather conditions and can be fifty percent larger during summertime. This figure is very important when designing the system, as we will see later.

DC Maximum Power Point Current (Impp)

The DC maximum power point current is the amperage that the solar module will output at standard test conditions in normal operation.   It is important to note that this value will vary a lot dependent on weather conditions and can be over fifty percent larger during summertime.

DC Rated System Voltage

This is very important design value. It is a rating of how many modules can be safely connected together in series. You should never exceed this voltage value when adjusting for the minimum and maximum temperatures of the area that the system is being installed. This value basically limits the number of modules that can be connected in series in the system.

3. Photovoltaics and Weather

The performance of any solar photovoltaic system is dependent on the weather. The main factors that affect the system performance are irradiance, temperature, shade, latitude and how dirty the solar modules are. Let's now explore the effects of the weather in more detail:

Irradiance

Irradiance is a measure of how much sunlight the solar module is receiving. It is given in watts per meter squared or W/m². Standard Test Conditions (STC) uses a value of 1,000 W/m². This value can range from 0 W/m² at night through to over 1,500 W/m² during a day interspersed with large fluffy clouds. This value of 1,500 W/m² is larger than what you would receive in space. The reason why we can get greater values at ground level is due to what is known as the "cloud effect". Normally the sunlight is traveling in a straight line from the sun to our solar module with some atmospheric scattering. However, when clouds are present they can also reflect and can act like lenses to send some extra sunlight onto the solar modules. This extra light is converted into extra energy and this is seen largely as an increase in power from the system. This effect can be a few minutes long in duration when it occurs.

The diagram on the next page demonstrates the "cloud effect".

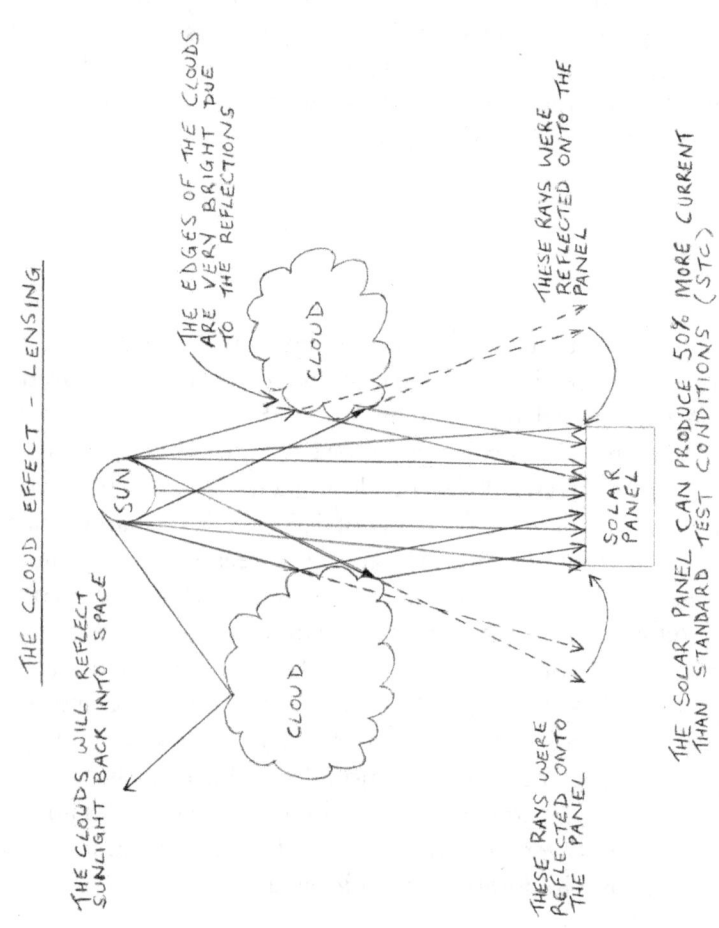

Other effects on irradiance are the snow effect, the lake/ocean effect and the building effect. Snow cover, water and glass covered buildings can reflect extra sunlight onto the solar power system. If you are installing a system in an area that has any of these, it is important to account for it. Each effect can produce an increase in power output.

In wintertime generally the system will operate at below the standard test conditions values and in summertime it will generally exceed these values. During designing the system you will need to see where your maximum power need is and perhaps increase the size of the system accordingly if it is in wintertime.

Air Mass

Air mass is a measurement of the the amount of atmosphere that the sunlight has to pass through to get to the ground. It varies with the seasons and also the location on the earth. Within the tropics, air mass will reach its maximum value of 1 during summertime. Airmass 1 corresponds to the sun being directly overhead, air mass increases as the sun moves from directly overhead down to the horizon.

All USA solar modules are rated for air mass 1.5 which corresponds to a central USA location. When in a southerly location you will approach air mass 1 which will increase power output by about 13% from STC in the USA.

Locations that are at or near air mass 1 in the USA are all Hawaiian islands, Florida and Texas. Approximately half of the USA is located between air mass 1 and air mass 1.5. If you are working on systems that are located in these Southern USA states, you will get more power out of these systems due to a decreased air mass.

Temperature

Temperature will affect the system to a much lesser extent than irradiance. The cooler the system is below 25 degrees Celcius, the more power it will produce. Correspondingly, the hotter the system is above 25 degrees Celcius, the less power it will produce. Temperature can affect solar photovoltaic systems power output by about twenty percent.

Shade

It is undesirable to shade solar modules as it can significantly affect the performance of the system. When studying the location of where to install a system, always factor in the surroundings for shading effects. Avoid shading with solar photovoltaic power systems.

Wind

Wind will provide cooling to the photovoltaic modules and it is an aid to power production. A breezy location will provide improved performance from the system. When mounting solar modules onto racking, it is good to allow spaces between the solar modules in order to aid with cooling airflow around the modules and also reduce wind resistance. When choosing solar modules and mounting systems, it is important to ensure that they a are rated for the wind speed of the area that you are installing them into.

Altitude

A higher altitude location will improve the amount of irradiance that the system will receive, due to less scattering and absorption of the sunlight by the atmosphere. It also acts as a natural cooler of the system which further improves system performance. Generally a high altitude location will have a higher percentage of clearer skies during a year which will give a higher energy yield from the system.

Snow and Ice

Snow and ice should not affect a solar module, other than obscuring its view of the sun. Tracking systems can be affected by this and in some snowy locations it is advisable to park the solar system facing South during these periods. The reflection from the snow will increase the power from the system in Winter time.

Hail

Hail can break solar modules, so it is important to know type of hail that your area can receive. If you get large golf ball size hail, you may not want to install glass solar modules. Solar modules are tested for hail and pass the tests even if the glass module breaks. The test just ensures that the modules remains intact when broken. Glass solar modules are hard to break and normal sized hail should have no effect.

Dirt

Clean solar modules are the desirable configuration for a system. However, dust and dirt will get onto the surface of the modules and will degrade performance by up to 10% on average. Cleaning the modules is very much a function of the location where they are installed and also how dirty they are. Most people will clean on an as needed basis, generally when they are visually very dirty. Always follow the manufacturers instructions for cleaning your particular modules and remember that solar modules are operating with electricity flowing in them when exposed to light. Night time cleaning is recommended for safety.

Lightning

Lightning can affect solar modules, especially on large systems that cover fields. Good equipment grounding is the way to deal with this threat. A low resistance ground will generally dissipate lightning away from a solar module that is struck by lightning. Generally, the damage should be limited to only the solar module that was struck. If a cable is struck, then lightning surge arrestors can limit the damage in the system. These are generally installed in the inverter and on larger systems, in combiner and recombiner boxes.

Seasons

We have four distinct seasons of Winter, Spring, Summer and Fall. We can word this another way as Winter Solstice,

Spring Equinox, Summer Solstice and Fall Equinox. What does this mean to a solar power system? Two things:

- The length of the day
- The angle of the sun (air mass)

Winter solstice is the shortest day of the year and summer solstice is the longest day of the year. Spring and fall equinoxes are when day time is the same length of time as night time.

Regarding the angle of the sun, Winter solstice is when the sun is at the lowest in the sky, or 23.5 degrees below the equator and Summer solstice is when it is 23.5 degrees above the equator. Spring and fall equinoxes are when the sun is on the equator.

For our solar power system, this means that we will produce our largest voltage in wintertime when it is the coldest and we will produce our largest current when it is summertime with peak irradiance.

Due Diligence

It is important when choosing your solar power generation system that you are aware of the annual climatic conditions to expect. Amongst the data that you should have is:

- Historic annual minimum temperature
- Historic annual maximum temperature

- Historic annual maximum wind speed
- Historic annual snow fall depth
- Historic annual hail size
- Historic annual peak irradiance
- Historic monthly irradiance

With these values you will be able to make educated engineering decisions regarding the design of your system.

4. System Selection

System selection is the most complicated stage of solar photovoltaic system designs and will determine how well your system works. When done well, your system will give years of trouble free operation.

Solar Modules

Solar modules typically represent the largest investment in the system. Traditionally silicon solar modules have been used, but now many new types of technologies are emerging such as thin film and so on. All solar modules are tested to the Underwriters Laboratory (UL) 1703 standard in the USA.

Silicon is the best understood, the most efficient and has been around for many decades. The newer film type solar modules are less efficient and cheaper to purchase. Unfortunately more thin film modules are needed to generate the same power as silicon, and this increases the system physical size and associated support systems such as cabling, racking, installation costs and so on.

Generally any decision on which technology to use is driven by market rates for each type of technology, aesthetics and personal preference. Solar modules are a commodity and their prices can fluctuate rapidly.

Mounting System

There are three ways to mount your modules:

- Fixed tilt
- Single axis tracker
- Dual axis tracker

All have their pros and cons.

The fixed tilt system is the most common and is widespread. The solar modules are either mounted to a roof, building or are ground mounted in a fixed position inclined to face south at a tilt angle matching the the latitude. Some systems allow you to adjust the tilt angle of the modules for the season, but it appears that most people prefer the low maintenance option of mounting the modules into a fixed position for the entire year. The fixed tilt system is the most reliable configuration and also the lowest cost. The downside is it has the lowest annual energy output of the mounting systems.

The single axis tracker works well in. The solar modules are mounted on a rotating North-South axis which allows them to track from East to West during the day. There are two types of single axis trackers generally available. The first has the North-South axis mounted horizontal and the modules can track in the East to West direction. This system works well in or near to the tropics where the sun can be almost directly overhead. The second has the North-South axis inclined to match the latitude and this enables the solar modules to face the sun in spring and fall.

This system works better as you move further away from the tropics. A single axis tracker can increase power output by about 25% when compared to a fixed tilt system. The single axis tracker does not cost much more than a fixed tilt system and the extra expense is generally offset by the extra annual energy yield of the system.

The dual axis tracker has the modules tracking the sun from sunrise to sunset, keeping the solar modules in the optimal position for maximum power generation. A dual axis tracker can increase annual energy output by about 40% when compared to a fixed tilt system. The downside to a dual axis tracker is that it requires a lot of space, can be very tall, has a complicated control system, they are expensive and they are the highest maintenance system.

The diagram on the next page shows the differences for each tracker system at noon with the seasons.

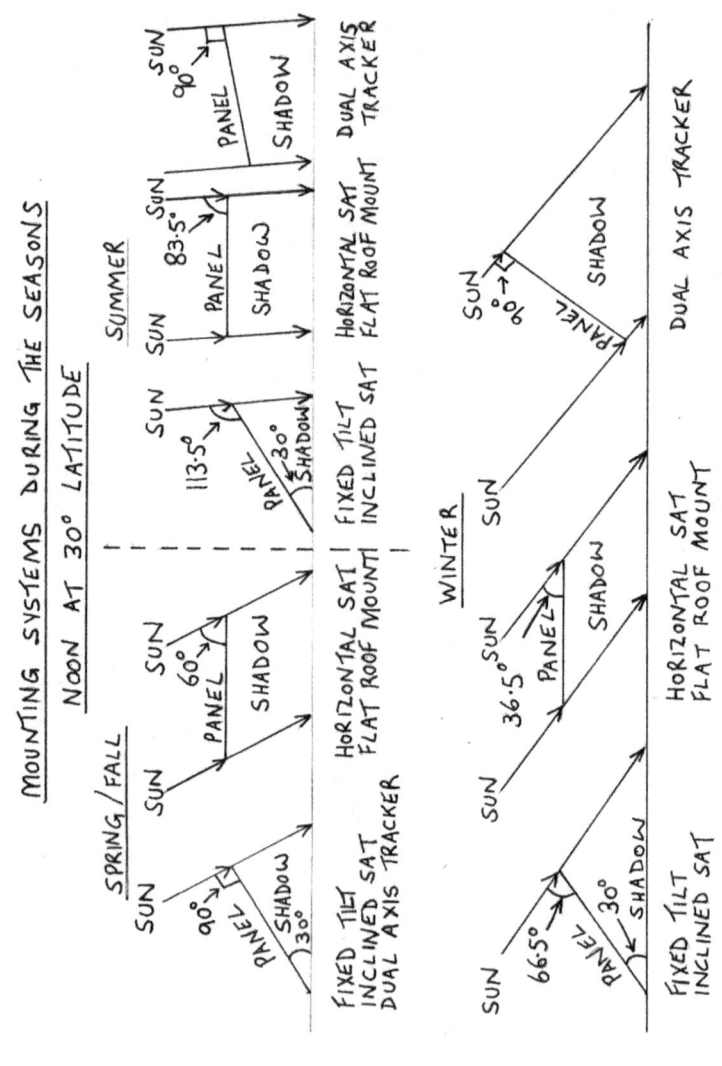

Cabling

For all outdoor cabling exposed to the sun, solar rated cable should be used due to its resistance to ultraviolet radiation. Once inside a building, conduit or underground this requirement need not apply.

On ground mounted installations it is important to remember that animals will be able to access the cables and equipment, so you will need be familiar with your local wildlife in the area. Where possible, enclose all accessible cabling with protective covers.

Conduit or ducting is recommended for all underground runs as it can be easily rewired to a larger size if you find that you system is generating more power than designed for or if a cable goes faulty.

Inverters

Inverters come in many different types and sizes. This book is dedicated to grid tie inverters and this is what we will consider. All residential and commercial solar photovoltaic grid tie inverters have to comply with UL1741 in the USA. This ensures that all grid tie inverters meet the requirements of the utility authority. Some of these requirements are below:

- Disconnect from grid during power cuts
- Protection against faults
- Good power quality output

Solar Design Hints and Tips

All DC Grid Connected Solar Photovoltaic Systems Need

- Equipment Ground.
- DC Disconnect.
- Inverter.
- AC Disconnect.

As They Get Larger They Also Need

- Combiner box.
- Recombiner boxes.
- Transformers and switchgear.
- Distribution and transmission systems.

Important Guidelines for Tracking Systems

- Try to put no more than one tracking system on one inverter.
- Install blocking diodes in each string to prevent reverse current flowing on multiple tracking systems installed on one inverter.

DC component selection

- Always make sure that DC rated components are used.

- Components must exceed specifications for maximum voltage.

- Components must exceed specifications for maximum continuous current.

- All outdoor cables must be solar rated.

- All outdoor connectors must be solar rated.

- De-rate equipment, cables and fuses for installed conditions such as:

  - High ambient enclosure temperatures.

  - Burial.

  - Enclosed conduits and ducts.

  - Heating effects.

  - Cooling effects.

  - High power cycling.

- Cables, breakers and fuses:

  - At least 156% larger than the short circuit solar photovoltaic current.

  - Increase current as needed for reflections.

  - Never exceed the interrupt ratings of the fuses and breakers.

- Use metal enclosures for effective heat dissipation from the electrical equipment.

AC Component Selection

- Always make sure that AC rated components are used.

- Components must exceed system specifications for maximum voltage.

- Components must exceed system specifications for maximum current.

- All outdoor cables must be solar rated.

- All outdoor connectors must be solar rated.

- De-rate equipment for installed conditions such as:

  - High ambient enclosure temperatures.

  - Burial.

  - Enclosed conduits and ducts.

  - Heating effects.

  - Cooling effects.

  - High power cycling.

- Cables, breakers and fuses:

  - At least 125% larger than the maximum AC current.

  - Apply de-rating for highest expected fuse ambient temperature per fuse manufacturers de-rating tables.

  - Never exceed the interrupt ratings of the fuses and breakers.

  - Use metal enclosures for effective heat dissipation.

Solar Photovoltaic Design by Steven Magee

Enclosure Mounting

- Exterior equipment locations should always be in the shade.

Inverters

- More inverters improves maximum power point tracking on the system
- Keep inverter power down to about 250 kW AC maximum per single inverter.
- Try to keep below 100 strings per inverter.
- If possible, mount inverters in shaded locations, consider constructing a shade canopy if needed.
- If inside a building or structure, ensure that the indoor ambient temperature can never exceed the inverter ambient temperature ratings.
- Use proven inverter technology for your system size.

Manufacturer Instructions

- Always design your system in accordance with the manufacturers installation manuals
- Follow the maintenance schedules in order to maintain warranty coverage

Codes

- Always design to local photovoltaic electrical codes.
- Always design to local building codes
- Consult with qualified engineers
- Have qualified engineers approve designs
- When in doubt, engineer on the side of safety.

Labeling

- Make sure all equipment is labeled in accordance with the local codes and the manufacturers instructions.

Safety

- Make sure all system and equipment grounding is installed correctly
- Make sure all covers are installed
- Make sure DC polarity of cabling is correct
- Make sure AC phase rotation is correct for three phase systems.
- Follow the local safety codes

System Inefficiencies

The system inefficiencies comprise of the following component de-rate factors:

- PV module nameplate DC rating 0.80 - 1.05
- Inverter and Transformer 0.88 – 0.98
- Mismatch 0.97 - 0.995
- Diodes and connections 0.99 – 0.997
- DC wiring 0.97 - 0.99
- AC wiring 0.98 – 0.993
- Soiling 0.30 - 0.995
- System availability 0.00 - 0.995
- Shading 0.00 – 1.00
- Sun-tracking 0.95 – 1.00
- Age 0.70 - 1.00

A well built solar photovoltaic system should be in the vicinity of 78% to 85% overall efficiency conversion from DC to AC at the point of interconnection.

System size

There are two ways that are traditionally used to quote system sizes

- Installed solar module DC power at STC

- Expected AC output at system interconnection at STC

It is important when designing a system that you quote both values accurately to the customer.

5. Grid Interconnection Requirements

Grid interconnection is different all over the world and it will depend on who's grid you are connecting into. Each utility will have its own requirements for grid interconnection and you will need to contact them to obtain their documentation on this.

In general they are looking for the following attributes:

- Address of installed system
- System generation capacity
- Power factor of system generation
- Inverter manufacturer(s) and model number(s)
- Sufficient interrupt ratings on breakers and fuses
- For residential and commercial installations only, confirmation of anti-islanding feature.
- Certification of country standards of manufacture of equipment.

It is important as a utility customer that you use equipment that will disconnect upon the grid power failing. This is to ensure that the grid is not back fed and energized from your equipment. This represents a significant hazard to utility line workers and many have been exposed to electrical hazards caused by this effect in the past. UL 1743 listed inverter equipment meets this requirement.

For a large utility scale system, you will want the utility to supply the following details:

- Interconnection fault current
- Interconnection voltage
- Transformer configuration
- Relay settings

With these values you will able able to design your utility interconnection.

6. Residential Design

Residential design for the purposes of this book is systems up to 10 kWp DC at STC. Most readers of this book will be working with this size of system. Residential systems represent the starting point for our designs in this book and underpin the larger system designs. To understand solar design, this is your starting point.

Residential systems are relatively easy to design electrically and have very few solar strings, sometimes they may only have one string. Something to consider with a residential system is the size of the solar modules that you will use. Since most solar systems on a residential system will be mounted to the homeowners roof, it is wise to select a solar module that is easily handled when up on the roof. Sometimes bigger is not better when installing roof mounted systems.

For this example we will assume that this will be a tiled, thirty degree slope, South facing roof mounted system. An assessment must be made to make sure that the roof is strong enough to mount the solar power system to. A structural engineer can perform this assessment.

Once confirmed that we can mount a solar photovoltaic system to the roof, you will need to identify where the roof supports or trusses run. You will want to attached your mountings to this location. A roofer can help you with removing the tiles and attaching your chosen mounting to the roof. Sometimes you will be able to drill the tile and refit it, other times you will want the roofer to flash the mounting with lead to waterproof the location in the absence of the tile being refitted.

The system that we will design will be a 5 kWp DC system which will be typical for most homes.

The solar module specifications are as follows:

- Maximum Power at STC = 235 W

- Tolerance of Power = +10% / -5%

- Open circuit voltage at STC = 37V

- Maximum power voltage at STC = 30V

- Short Circuit Current at STC = 8.6A

- Maximum power current at STC = 7.84A

- Maximum system voltage = 600VDC

- Series fuse rating = 15A

- Power temperature coefficient = -0.485%/°C

- Voltage temperature coefficient = -0.351%/°C

- Current temperature coefficient = 0.053%/°C

- Normal operating cell temperature (NOCT) = 47.5 °C

- Size (L x W x H): 48 in x 36 in x 1 in

Our Inverter specifications are:

DC Specifications:-

- Continuous Power @ 240 VAC: 5150W

- Recommended Max PV (STC): 6200W
- MPPT Voltage Range: 200V - 550V
- Maximum Input Voltage: 600 VDC
- Strike Voltage: 235 VDC
- Maximum Input Current: 25A
- Maximum Input Short Circuit Current: 30A
- Fused Inputs: 4

AC Specifications:-

- Continuous Power @ 240 VAC: 4900W
- Voltage Range @ 240 VAC: 211-264VAC
- Frequency Range: 59.3-60.5Hz
- Continuous Current: 20.7A
- Output Current Protection: 30A
- Max Backfeed Current to PV: 0A
- Power Factor: Unity >0.99
- Total Harmonic Distortion: <3%
- Efficiency: 96%

General:-

- Enclosure: Rainproof, NEMA 3R
- Housing Material: Painted Aluminum
- Ambient Temperature Range: -25°C to +55°C
- Weight: 61.7 lbs, 28Kg
- Cooling: Convection & Fan Assist

- Wire Sizes: 12 to 6 AWG input & output connections

- Size (L x W x H): 28 4/5 in x 17 ¾ in x 8 ¼ in (732mm x 454mm x 210mm)

- Standards: UL1741/IEEE1547, IEEE1547.1, ANSI62.41.2, FCC Part 15B

- Warranty: 15 Years

Location Area Specifications are:

- Historic annual minimum temperature: -10°C

- Historic annual maximum temperature: +40°C

- Historic annual maximum wind speed: 80 MPH

- Historic annual snow fall depth: None

- Historic annual hail size: None

- Latitude: 30

- Reflections: None

Of note is how the solar module ratings vary with temperature. Voltage is affected the most with current almost unaffected.

This solar modules has a normal operating cell temperature (NOCT) of 47.5 degrees Celsius. This is 22.5 degrees above the STC rating. This gives:

- NOCT Power: 210 W

- NOCT Voltage: 34.1 V

- NOCT Current: 8.7 A

As we can see, our solar module looks quite different at a higher temperature than STC. This is an important concept to grasp in solar photovoltaic design.

It is not just the solar modules that are affected by temperature. Our equipment such as terminals, cables, fuses and so on are affected also. As such, we will use the following values for our system design de-ratings for this area:

- 80°C Attic temperature and solar module temperatures
- 60°C Enclosure temperature
- 40°C Underground temperature

Now onto the design of our system:

To get a 5 kWp DC system at STC, we need to divide 5 kW by STC power:

5,000W / 235W = 22 modules minimum

We need to put our strings together for a voltage below 600 VDC. This requires an adjustment to the open circuit voltage to allowing for a minimum temperature for this system of -10 Celsius

Temperature difference = STC – Lowest temperatures

$=25°C - (-10°C) = 35°C$

Open circuit voltage adjustment $= -35°C$ x (-0.351%)

$= 12.285\%$ adjustment

Solar module maximum STC voltage $= 37V$

Adjusted solar module maximum voltage $= 37V * 12.285\%$

$= 41.55V$

We call a group of solar modules electrically connected together in series a "string". We do this in order to increase the DC voltage to a level that the inverter can use. So our maximum number of solar modules in a string is:

$=$ System voltage / adjusted solar module maximum voltage

$= 600V / 41.55V$

$= 14$ solar modules maximum in strings

Our solar module is 235 Watts, so this gives a string wattage at STC of:

String Wattage $=$ Number of solar modules x module watts

$= 14 \times 235W = 3,290W$

String Voltage = Number of solar modules x adjusted voltage

$= 14 \times 41.55V = 581.7V$

Number of strings in system = System watts / string watts

$= 5,000W / 3,290W = 1.5$

We need to round up to make our strings equal, so we would go to two strings at this point:

$5,000W / 2 = 2,500W$

Each string needs to be at least 2,500W. We divide this figure by our module wattage to get our string number:

$2,500W / 235W = 10.63$ modules

We round up to 11 modules in each string to give:

String watts = 11 modules x 235W = 2585W

Two strings = 2 Strings x 2585W = 5170 Watts

String maximum voltage = adjusted maximum voltage x number of modules in string

= 41.55V * 11 = 457V

Solar modules generate their maximum voltage at their lowest temperature. Conversely, they generate their minimum voltage at their highest temperature.

Assuming that the maximum temperature that the solar module will operate at is 40°C higher than ambient temperature:

Solar module adjusted minimum voltage = STC MPPT voltage - ((Annual maximum temperature + 40°C increased module temperature adjustment - 25°C STC temperature) x Voltage temperature adjustment percentage x STC MPPT voltage)

= 30V + ((40°C + 40°C - 25°C) x (-0.351%) x 30V)

= 24.2V

String minimum voltage = number of strings x solar module adjusted minimum voltage

= 11 x 24.2V

=266.2V

Looking at our minimum and maximum values and comparing the to the inverter DC voltage specifications, both values will work well with this inverter.

266.2V Minimum string voltage > 235V inverter strike DC voltage

457V Maximum string voltage < 600V Maximum input DC voltage

Next we look at our maximum current for the string. From the data sheet for the solar module we see that the solar module short circuit current at STC is 8.6A with an adjustment for current temperature coefficient = 0.053%/C. Now let's see how much current we should expect from the system at our maximum ambient temperature of 40°C.

Adjusted maximum solar module current output = Short circuit current at STC + ((annual maximum temperature + 40°C increased module temperature adjustment - 25°C STC temperature) x current temperature coefficient x STC short circuit current)

= 8.6A + ((40°C + 40°C – 25°C) x 0.053% x 8.6A

= 8.85A

As you can see, there is very little effect on current by the increased temperature of the module.

We now calculate the maximum continuous circuit current in NEC690.8(A) which gives:

Maximum continuous circuit current = Short circuit current x 125%

= 8.6A x 125%

= 10.75A

The National Electric Code requires the fuses and cables to be rated at least 125% higher than the maximum continuous circuit current. Per NEC690.8(B) this gives:

Overcurrent device ratings = Maximum continuous circuit current x 125%

= 10.75A x 125%

= 13.44A

The National Electric Code includes this 25% factor for fuse and cable de-rating. We will use the 15A maximum fuse as recommended by the manufacturer.

String fuses =15A

This is higher than needed and incorporates sufficient de-rating for the fuse for this application.

The cable amperage size should be the same or larger than the fuse size. Since our system is mounted to the roof and our cables will pass through the attic in free air we will use NEC table 310.16 to obtain our cable size. Since the attic is a hot location, we will de-rate the cable for a maximum temperature of 80°C.

Cable current =>15A

80°C de-rating = 0.41

De-rated cable size = fuse size / 80°C de-rating

= 15A / 0.41

= 36.6A

Looking at NEC table 310.16 we see that 10 AWG cable exceeds this at 40 amps.

String cable size = 90°C 10 AWG

Finally, our electrical ground conductors are sized. There are two ground conductors used, one for the system and one for the equipment.

The system ground is achieved by connecting one conductor of our two wire DC system to ground. In this particular case, this is achieved internally inside the inverter using the negative conductor.

The equipment ground is to protect all exposed non current carrying metal parts in the system, such as module frames. When grounding the module frames, use the recommended grounding lug that the manufacturer specifies in their installation manual. NEC 690.45 specifies the size of the ground to be 125% of the total photovoltaic originated source current of the circuit. In our case this is:

Solar module short circuit current = 8.6A

Equipment ground current = solar module short circuit current x 125%

= 8.6A x 125%

= 10.75A

From NEC table 310.16 we can obtain our equipment ground cable size after de-rating.

de-rated ground cable size = Cable current / 80°C de-rating

= 10.75A / 0.41

= 26.22A

Looking at NEC table 310.16 we see that 90°C 12 AWG copper cable exceeds this at 30 amps.

Equipment ground cable size = 12 AWG

For each individual string we will use a 12 AWG cable to our ground location at the inverter.

The support racking for the solar modules will be grounded in accordance with the manufacturers instructions and local codes.

Using voltage drop tables the voltage drops are less than 3% in this application since there is only 100 feet of cable between the solar modules and the inverter.

Since the distance from the solar modules to the inverter is so short, we will use the inverter inputs to combine the two solar string circuits. We will install a DC disconnect switch at the nearest accessible location to the solar panels.

Now onto the AC circuit:

Our connection point to the grid will be at the utility meter. This eliminates any need to verify if the fuse board is capable of being back fed by the solar photovoltaic system. Our inverter will be mounted next to the meter with a AC fused disconnect switch between them. The fuses will be sized according to the inverter output fusing specification of 30A. Our cable needs to be sized to at least 30 amps also.

Cable size = fuse size x 60°C cable de-rating

= 30A / 0.71

= 43A

Looking at NEC table 310.16, we see that 90°C 8 AWG cable will meet this need at 55A.

Equipment ground cable = inverter AC current / 60°C de-rating

= 21A / 0.71

= 30A

Looking at NEC table 310.16, we see that 90°C 12 AWG cable will meet this need at 30A.

The annual maximum wind speed is 80 MPH and a solar module mounting system will be used that exceeds this requirement.

So here are our system specifications:

- Inverter: 5 kW, single phase, 240 VAC, 600 VDC
- Inverter 240VAC cabling: 90°C 8 AWG copper
- Inverter AC ground cable: 90°C 12 AWG copper

- Solar string cabling: 90°C 10 AWG
- Solar string equipment ground cable: 90°C 12 AWG copper
- Solar string fuses: 15A
- Number of solar string fuses: 2
- Number of solar strings: 2
- Number of solar modules: 22

Have your solar photovoltaic design checked by your local solar photovoltaic licensed electrician before submitting the plans for approval to you local authority.

The system can now be drawn and is shown on the following page.

Solar Photovoltaic Design by Steven Magee

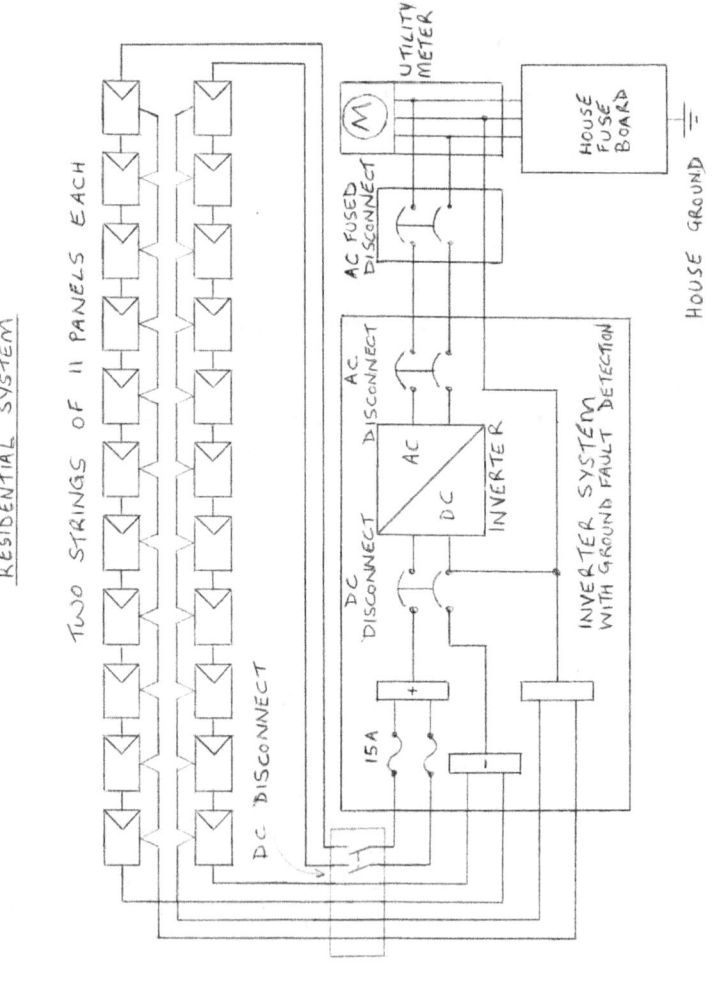

7. Commercial Design

Commercial design for the purposes of this book is systems from 10 kWp DC to 1 MWp DC at STC.

New ideas in this section are

- MPPT tracking performance
- Combiner boxes

MPPT Tracking Performance

Maximum power point tracking (MPPT) is at its most efficient on a single solar module string. As you increase your system size and add more strings to it, it can reduce the performance of MPPT. The MPPT tracking will only be as good as the worst performing strings connected to it. As such it is important to limit the number of strings connected to an inverter. I would suggest that no more than one hundred strings be connected to an inverter for this reason.

Combiner boxes

These are needed to connect our strings together so that they can be combined into a larger cable. They generally come in 6, 12, 18 and 24 string capacities. More advanced models have the capacity to monitor each string connected to it using a computer network and the associated software. Each string will have a fuse or circuit breaker for it.

250 kW System Design

Our commercial system will be a single 250 kW inverter design that will connect into the distribution system of the commercial facility.

We will use the same solar module as before and this time we will increase the number of modules in the string to the maximum voltage that the inverter can accept.

This will be a ground mounted, 30 degree fixed tilt system that connects into the commercial facility electrical system at 480 volt, three phase grounded system. We have verified that the existing facility switchgear is capable of being back fed by the solar photovoltaic system and that the bus system can handle the current passing through it.

Inverter Specifications

Our Inverter specifications are :

Input Parameters:-

- PV array configuration: Negative ground
- MPPT Voltage Range: 320V - 600VDC
- Maximum Input Voltage: 600 VDC
- Maximum Input Current: 814A

Output Parameters:-

- Maximum Continuous Output Power: 250kW (250kVA)
- Voltage Range @ 480 VAC: 422 – 528 VAC
- Nominal Voltage: 480 VAC
- AC Voltage Range: -12%/+10%
- Frequency Range: 59.3-60.5Hz
- Nominal Output Frequency: 60 Hz
- Number of Phases: 3
- Maximum Output Current Per Phase: 301A
- Power Factor at Full Load: >0.99
- Total Harmonic Distortion: <3%
- Efficiency: 96%

Temperature:-

- Operating Ambient Temperature Range (Full Power): -20°C to +55°C
- Storage Temperature Range: -30°C to +70°C
- Cooling: Forced Air

Noise:-

- Noise Level: <65dB(A)

Combiner:-

- Number of Inputs and Fuse Rating: 10 (160ADC)
- Enclosure Rating: NEMA 3R, IP 44
- Enclosure Finish: RAL-7032
- Inverter Cabinet Dimensions (Height x Width x Depth): 92.6" x 117.7" x 43.3" (235.2mm x 298.96mm x 109.98mm)
- Inverter Cabinet Weight: 4500 lbs (2,041 kg)

Testing Certifications:-

- Standards: UL1741, CSA 107.1-01, IEEE1547, IEEE C62.41.2, IEEE C62.45, IEEE C37.90.1, IEEE C37.90.2
- UBC Zone 4 Seismic Rating

Warranty:-

- Five Years

Monitoring:-

- Third Party Compatible

To get a 250 kWp DC system at STC, we need to divide 250 kW by STC power:

250,000W / 235W = 1,064 solar modules minimum

We need to put our strings together for a voltage below 600 VDC. This requires an adjustment to the open circuit voltage to allow for a minimum temperature for this system of -10 Celsius.

Temperature difference = STC – lowest ambient temperature

=25°C - (-10°C) = 35°C

Adjustment: -35°C x (-0.351%) = 12.285% adjustment

Solar module maximum STC voltage = 37V = Voc

Adjusted solar module maximum voltage = 37V x 12.285%

= 41.55V

Maximum number of solar modules in string = inverter maximum DC voltage / adjusted solar module maximum voltage

= 600V / 41.55V

= 14 solar modules in strings

Our solar module is 235 Watts, so this gives a string wattage at STC of:

String wattage = number of solar modules x module watts

= 14 x 235W = 3,290W

String voltage = number of solar modules x adjusted voltage

= 14 x 41.55V = 581.7V

Number of strings in system = system watts / string watts

= 250,000W/ 3,290W = 76 strings

Total DC power = 76 Strings x 3290 W = 250,040 Watts

String maximum voltage = adjusted maximum voltage x number of modules in string

= 41.55V x 14 = 581.7V

Solar modules generate their maximum voltage at their lowest temperature. Conversely, they generate their minimum voltage at their maximum temperature.

Assuming that the maximum temperature that the solar module will operate at is 40°C higher than ambient temperature:

Solar module adjusted minimum voltage = STC MPPT voltage - ((Annual maximum temperature + 40°C increased module temperature adjustment - 25°C STC temperature) x

Voltage temperature adjustment percentage x STC MPPT voltage))

= 30V+((40°C + 40°C - 25°C) x (-0.351%) x 30V)

=24.2V

String minimum voltage = number of strings x solar module adjusted minimum voltage

= 14 x 24.2V

=338.8V

Looking at our minimum and maximum values and comparing the to the inverter DC voltage specifications, both values will work well with this inverter.

338V Minimum string voltage > 320V inverter minimum DC voltage

581.7V Maximum string voltage < 600V inverter maximum input DC voltage

We now adjust for maximum continuous current using NEC 690.8(A) which gives:

Normal circuit current = maximum solar module short circuit current output x 125%

= 8.6A x 125%

= 10.75A

The National Electric Code 690.8(B) requires the fuses and cables to be rated at least 125% higher than our normal circuit current. This gives:

NEC circuit sizing and current = 10.75A x 125%

= 13.44A

This is the factor for fuse de-rating. If the manufacturer specifies fuse de-ratings for certain temperatures, then this value should be calculated and the larger value of the two calculations should be used.

Using the solar module manufacturers 15A maximum fuse is what we will select for this application.

String fuses =15A.

The cable size should be the same or larger than the fuse size, as this is what the fuse protects. Since our system is ground mounted and our cables will pass through raceways we will use NEC table 310.16 to obtain our cable size. The conduits are in a high ambient location and we will de-rate the cables for a maximum ambient temperature of 60°C.

Cable current => 15A fuse size

60°C de-rating = 0.71

De-rated cable size = Cable current / 60°C de-rating

= 15A / 0.71

= 21.12A

Looking at NEC table 310.16 we see that 90°C, 14 AWG copper cable exceeds this at 25 amps.

Circuit Cable size = 90°C 14 AWG copper cable

Finally, our electrical ground conductors are sized. There are two ground conductors used, one for the system and one for the equipment.

The system ground is achieved by connecting one conductor of our two wire DC system to ground. In this particular case, this is achieved internally inside the inverter. The negative cable is grounded when the system is in operation.

The equipment ground is to protect all exposed non current carrying metal parts in the system, such as module frames. When grounding the solar module frames, use the recommended grounding lug that the manufacturer specifies in their installation manual. NEC 690.45 specifies the size of the ground to be 125% of the total photovoltaic originated source current of the circuit. In our case this is:

Equipment ground current = Short circuit current output x 125%

= 8.6A x 125%

= 10.75A

From NEC table 310.16 we can obtain our equipment ground cable size after de-rating.

De-rated cable size = ground cable current / 60°C de-rating

= 10.75A / 0.71

= 13.43A

Looking at NEC table 310.16 we see that 90°C, 18 AWG copper cable exceeds this at 18 amps. However our minimum ground cable size per NEC table 250.122 is 14 AWG, so this is what we will use.

Equipment ground cable size = 90°C 14 AWG copper cable

We will use the ground clamps that the manufacturer specifies for the solar modules and the ground will be a continuous unbroken cable.

The mounting structure will be grounded in accordance with the manufacturers instructions and local codes.

We will now combine the solar strings into a larger cable using the combiner boxes. The outputs of the combiners will pass underground to the inverter.

Continuous current output of 10 string combiner:

10 Combiner = 10 strings x short circuit current x 125%

= 10 x 8.6A x 125%

= 107.5A

Our fuse and cable size will be further multiple of 125% per NEC 690.8(B)

Fuse size = continuous current output x 125%

= 107.5A x 125%

= 135A

From NEC 240.6(A) we see that the next standard fuse size is 150A and we will use this.

We match our cable size with the fuse size and de-rate it for a underground temperature of 40°C:

10 combiner de-rated cable size = fuse size / 40°C de-rating

= 150A / 0.91

= 165A

Looking at NEC table 310.16, we see that a copper cable size of 90°C, AWG 1/0 has a rating of 170A and is suitable for this application.

10 combiner box cable size = 90°C 1/0 AWG copper

Now for the ground conductor:

Ground conductor current = 10 strings x short circuit current x 125%

= 10 x 8.6A x 125%

= 107.5A

Now we de-rate this for an underground ambient temperature of 40°C:

de-rated ground conductor = ground conductor current x 40°C temperature de-rating

= 107.5A / 0.91

= 119A

From NEC table 310.16 we see that a 90°C 2 AWG copper cable is rated at 130 amps and this is what we will use.

Now let's do the same for the 9 combiner box:

9 Combiner continuous current = 9 strings x short circuit current x 125%

= 9 x 8.6A x 125%

= 96.75A

9 combiner fuse size = 96.75A x 125%

= 121A

NEC 240.6(A) next standard fuse size is 125A and this is what we will use.

9 combiner de-rated cable = combiner fuse / 40°C derating

= 125A / 0.91

= 138A

Looking at NEC table 310.16, we see that a copper cable size of 90°C 1 AWG has a rating of 150A and is suitable for this application.

9 combiner box cable size = 90°C 1 AWG copper

Now for the ground conductor:

Ground conductor current = 9 strings x short circuit current x 125%

= 9 x 8.6A x 125%

= 96.8A

Now we de-rate this for an underground ambient temperature of 40°C:

De-rated ground conductor = ground conductor current x 40°C temperature de-rating

= 96.8A / 0.91

= 107A

From NEC table 310.16 we see that a 90°C 3 AWG copper cable is rated at 110 amps and this is what we will use.

Now onto the AC circuit:

Our connection point to the grid will be at the utility meter distribution center. We will use a spare distribution switch to feed our inverter which will be located 100 feet away. The fuse size for this switch will be:

Inverter AC Fuses = (Inverter power x 125%) / sqrt(3) x inverter AC voltage

= 250 kW x 125% / sqrt(3) x 480V

= 376A

Using NEC 240.6(A) we see that our next standard size fuse is 400A and this is what we will use. If the fuse manufacturer supplies fuse temperature de-ratings, calculate this value and use the largest of the two values for the fuse size.

Our cable size will be 400A with a de-rating for 40°C for being underground.

Cable size = fuse size / cable temperature de-rating

= 400A / 0.91

= 440 amps

From NEC table 310.16, we see that 90°C 600 kcmil copper cable meets this requirement at 475 amps.

Now for the ground conductor:

Ground conductor current = AC circuit current x 40°C de-rating

= 376A / 0.91

= 414A

From NEC table 310.16 we see that a 90°C 500 kcmil copper cable is rated at 430 amps and this is what we will use.

The inverter will be grounded in accordance with the manufacturers instructions and local codes.

Looking at our annual maximum windspeed of 80 MPH, a solar module mounting system will be used that exceeds this requirement. The mounting system will be grounded in accordance with the manufacturers instructions and local codes.

The design that we are looking for is to be a long to run the length of the commercial premises on the South side. We are located at 30 degrees latitude so the solar modules will be tilted at 30 degrees, facing south.

4 divides well into 76 strings and this will give a design of:

76 strings = 4 rows x 19 strings

Looking into sizing this, our solar modules will be 3 feet wide. We will leave a 1 inch gap between each module for ventilation and expansion.

Row length = (number of strings in row x number of solar modules in a string x (solar module width + 1 / 12)) - 1/12

= (19 x 14 x (3 + 1/ 12)) - 1/12

= 820.17 feet long

We will place our inverter to the north of the modules to prevent shading from it and in the center of the row to minimize long cable runs. To the East on each row we will have ten strings and to the West we will have nine strings. Our combiner boxes will be 4 combiners of 10 strings and 4 combiners of 9 strings.

We need to make sure that our shading from each row is not affecting the row behind. To do this we use the equation:

Row Distance = sin (sun elevation + solar module tilt) x solar module length / sin (sun elevation)

We want our solar modules to function without shading after the sun is above 5 degrees of the horizon. Our solar module is 4 feet long and is mounted at 30 degrees tilt. This gives:

Row Distance = sin (5 + 30) x 4' / sin 5

= 26.33 feet

This measurement is from the front of one solar module to the front of the next.

We have three spaces between our four rows of solar modules, so we will cover

Total separation distance = number of spaces between rows x row separation distance

= 3 x 26.33 feet

= 79 feet

On this system we will have some long cable runs, some up to 500 feet from the inverter location and we will need to check on the voltage drop for the solar modules that are furthest out from the inverter. Using cable voltage drop tables, all cable sizes in this design are within the 3% voltage drop allowed.

So here are our system specifications:

- Inverter: 250 kW, 3 phase, 480 VAC, 600 VDC
- Inverter 480VAC cabling: 600 kcmil 90°C copper
- Inverter equipment ground cabling: 500 kcmil 90°C copper
- Inverter DC cabling for 10 input combiner: 90°C 1/0 AWG copper
- Inverter equipment ground cabling for 10 input combiner: 90°C 2 AWG copper
- Number of 10 input combiner boxes: 4
- Inverter DC cabling for 9 input combiner: 90°C 1 AWG copper
- Inverter equipment ground cabling for 9 input combiner: 90°C 3 AWG copper
- Number of 9 input combiner boxes: 4
- Solar string cabling: 90°C 14 AWG copper
- Solar string fuses: 15A
- Number of solar string fuses: 76
- Number of solar strings: 76
- Number of solar modules: 1,064
- Row Spacing: 26.7 feet

When the utility provided the interconnection details you will be able to calculate the fault currents in the design. With these values the correct switchgear can be selected, ensuring that it is rated for back feed.

Once the design is completed, it will need to be stamped by a licensed Professional Engineer to verify the design and the calculations.

The system can now be drawn and is shown in the following pages.

Solar Photovoltaic Design by Steven Magee

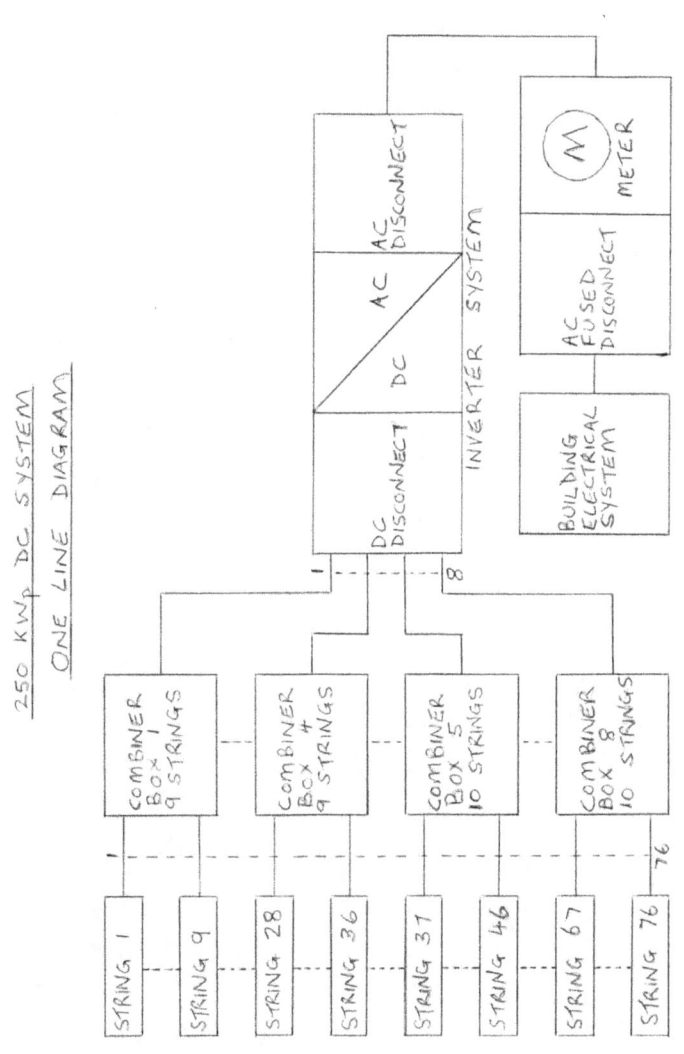

250 KWp DC SYSTEM
ONE LINE DIAGRAM

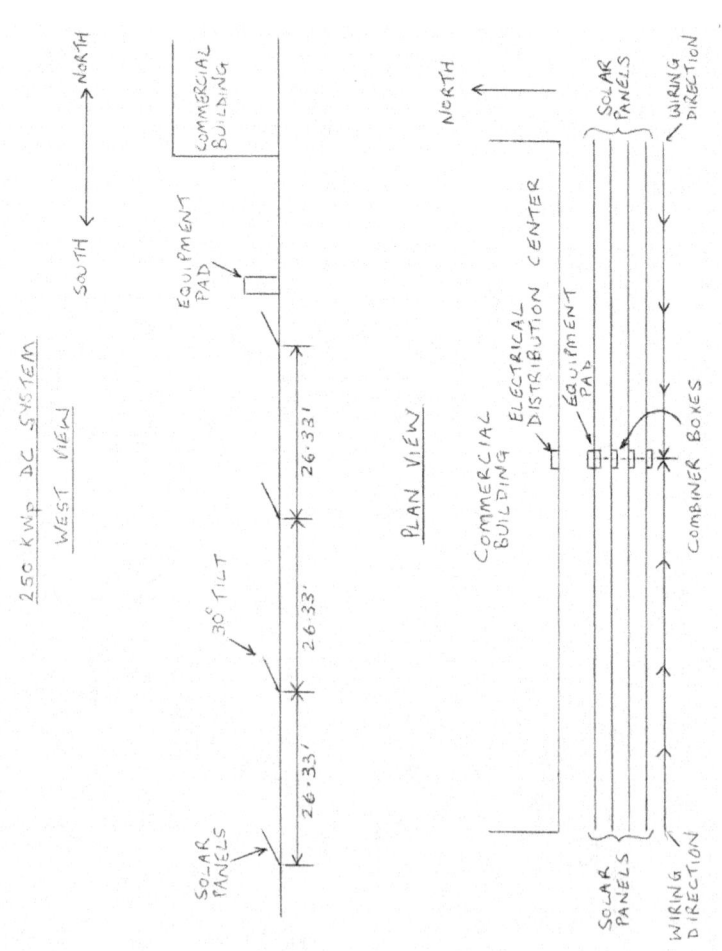

8. Utility Design

Utility design for the purposes of this book is systems over 1 MWp DC.

A new set of guidelines comes into play with these systems:

- Power Factor
- AC and DC Interrupt Current
- High Voltage Transmission
- Ride Through Capabilities
- Harmonics Requirements
- Large Power Swings
- Distributed or Centralized Generation
- Re-combiner Boxes
- National Electric Safety Code (NESC)
- 1,000 Volt DC Systems

Power factor

When generating power from an inverter it is generally set to unity. However, for a utility grid, this will cause problems with the power factor regulation on the grid. Feeding in a large amount of unity power factor energy into a grid that is operating at 0.9 lagging power factor will bring the overall power factor down to a lower level, such as 0.8 lagging. It is undesirable to operate the grid at this level and as such, control of the power factor being generated by the inverter(s) is desirable. Generally utility inverters can output their power factor between 0.9 leading and 0.9 lagging. Since the grid operates at 0.9 lagging due to large inductive loads, it is suggested that any utility inverter be initially set to generate at a power factor of 0.9 lagging.

A nice feature of utility inverters is that the power factor can be controlled. This offers the ability of power factor correction for the grid. When designing a utility system, it is important to design in the ability to control power factor for the utility authority.

AC and DC Interrupt Currents

At the utility level, this becomes a major concern. All fuses and circuit breakers have interrupt ratings. It is important when working with interrupt currents that you ensure that in the DC circuit that you are using the DC interrupt current ratings and in the AC circuit that you are using the AC interrupt current ratings. Generally, interrupt currents are in the range of tens of thousands of amps.

If you are using a large inverter, then it may have an equally large capacitor inside of it. It is important to find out the short circuit fault current for the inverter DC bus as you will need this to verify the interrupt current of your design. My recommendation is to use a DC design of 250 kilowatts STC or below to keep the DC fault currents to a manageable level.

You will need to ensure that you do not exceed the interrupt current of the fuses and circuit breakers used in the system. Sources of current that will need to be considered in the DC circuit are:

- Solar module maximum circuit current multiplied by the number of solar module strings connected to the inverter.
- Cable capacitance
- Inverter capacitance

In a inverter system of over a megawatt with several hundred strings spread over acres of land, these interrupt currents could be in the tens of thousands of amps range. Be careful when designing such large DC circuits and pay close attention to the DC interrupt current.

Long cable runs that are present on the DC circuits will generate capacitance effects. The cable manufacturers data sheet will specify the capacitance per unit length of the cabling and the total capacitance can be calculated for the cabling.

In the AC circuit, the largest fault current will generally be that from the grid. Before selecting your AC equipment, you will need to establish this value from the utility and calculate the fault currents for your particular design.

These are sources of fault currents in the AC circuits:

- The grid

- The solar power inverter systems
- Cable capacitance
- Power factor control capacitance
- Inductance

Make sure that any AC fault analysis includes all sources of power feeding into the fault.

Always calculate the fault currents for the DC and AC circuits and ensure all fuses and breakers can interrupt these currents.

High Voltage Transmission

Transmission for the utility grid can be at over one million volts. The previous designs that have been discussed operate at low voltages. Transforming the voltage up from this to medium and high voltage is straight forward.

You will need to know the following items to design the transmission interconnection:

- Interconnection voltage.
- Star or delta feed into the transmission network.
- Interconnection short circuit fault current.
- Transformer impedance.
- Harmonics requirements.
- Protection relay system required by the utility.

- Protection relay settings required by the utility.

With these, you will be able to design the solar photovoltaic system interconnection. It is important when designing the interconnection that you use electrical switchgear that is capable of being back fed in your design.

Ride Through Capability

Utilities want to generate power and if the grid voltage dips, they want their power generations systems to stay on line and "ride through" the dip. This is a different concept than previously discussed and is unique to the utilities.

It is important when building a system for a utility that you select a utility grade inverter that has this feature. The utility wants to keep power feeding into the electrical grid and wants the grid to stay on line. Generally the ride through feature can be set to keep feeding into the grid for a specified time before the inverter will shutdown. Check with the utility what there requirements are for this feature.

Harmonics Requirements

Most utility inverters generate power at below 3% harmonic distortion. This is generally acceptable by the utilities and multiple inverter systems can keep the harmonic distortion below this level. When selecting your inverter system, it is always good to get confirmation from the manufacturer that a large, multiple inverter system can keep within this tolerance.

Large Power Swings

Any solar photovoltaic system will suffer from large power swings that are proportional to irradiance. Solar cells directly convert irradiance into energy in real time. The main thing that can significantly effect the power output quickly are clouds. Clouds can increase the power as well as decrease the power output. Large, dense, fluffy broken clouds will reflect light from the sun and increase the power output from the solar modules that are in direct sunlight. As the cloud starts to approach the sun, lensing will occur that will focus more sunlight into the solar module. This surge in power can be a ramping up of normal power levels as higher than 25% increase for a few minutes. As the cloud progressively covers the sun, the power will start to drop off significantly with drops of over 90% from normal power levels. On a broken cloud day, these large power swings can occur within several seconds if the cloud is fast moving or over several minutes if they are moving slowly.

As you can imagine, these large power swings cause havoc with the grid and can look like faults to the grid control system. On a large power generation system, it is important to build into the grid control system feedback from the solar power system control system so that the grid can discern between a genuine fault and clouds passing over the solar power generation system.

The grid needs to be able to tolerate these large swings and alternate conventional power generation needs to be able to rapidly react to the power surges that the solar power system will generate during broken cloud days.

Distributed or Centralized Generation

Distributed power generation refers to when the solar power generation system is broken into many small parts and spread over a wide area. Residential installations and commercial installations would fall into this category. Distributed power generation is less prone to broken cloud effects due to the solar modules being spread over a large area. On a broken cloud day, some solar power systems will be in the sun while others will be in the shade. It creates an averaging of the broken cloud effect on the grid and is much more desirable.

Centralized power generation is when a large solar power generation system is installed in one location. Utility scale installations would be in this category. There are advantages regarding maintenance costs to centralize the utility solar power plant, but these seem to be negated by the cloud effects on the solar power system output. Centralized power generation works well when complimented by another source of fast response power generation, such as a gas turbine power generation plant. Thus the solar power generation system is used to offset the fuel consumption of the conventional power generation plant.

Recombiner boxes

Recombiner boxes are used to combine the outputs of two or more combiner boxes into a larger cable. They generally will have two to four large circuit breakers inside them, depending on the amount of combiner box circuits being combined.

National Electric Safety Code (NESC)

Now that we are on the utility side of the meter, a new standard applies. The National Electric Safety Code (NESC) covers the utilities. One thing that you will notice is that there is no mention of solar photovoltaics in the National Electric Safety Code. Due to this, we obtain our solar photovoltaics design equations from the National Electric Code (NEC). The National Electric Safety Code (NESC) and the National Electric Code (NEC) should be read in conjunction with each other. You will find on utility scale installations that you will refer to the two code books frequently.

1,000 Volt DC Systems

There has been a move towards 1,000 volt DC systems in the USA. 1,000 Volt systems are widespread in Europe. The advantage is that the efficiency of the DC circuit is increased due to lower power losses in the circuit. Cabling and equipment costs are reduced as less is used. I personally do not recommend these systems as a solar photovoltaic systems need maintenance at least annually and this high voltage increases the electrocution risks for the workers performing this maintenance. National Electric Code section 490.2 defines high voltage as over 600 volts. As such the provisions in National Electric Code section 490 as well as the National Electric Safety Code (NESC) should be followed.

10 MWp DC System

We will now design our utility scale system. For this design we will use 250 kW inverters. Our DC circuit, the mounting system and inverter design is the same as the commercial design from the previous chapter. We now have to connect 40 inverters together and feed the power into the grid. This is a relatively straight forward task and can be done in many ways, all of them would be equally suitable.

This particular design will consist of four 2.5 MW transformers feeding into distribution centers that connect to ten inverters each. This gives us a segmented design that during a large scale equipment failure, such as a faulty transformer, we should only see 25% of the power drop off from the grid.

Our commercial solar modules and inverter system are suitable for this design and we will now use that again here for each of our 250 kW systems.

We will end up with a grid of solar power systems on this design covering about 100 acres of land. We will build four blocks of solar, each a 2.5 MVA system.

We will take the outputs from each of the four systems, combine them using a medium voltage distribution center and the utility will connect them into the 24,000 volt utility poles that pass along the edge our piece of land.

Dry type, high efficiency power transformers are generally used in solar power applications. These have efficiencies of over 99% and are low maintenance. We will use these for our application.

Both our AC low (480V) and medium voltage (24,000V) circuits will be protected by ground fault detection systems. These will use protective relays to detect ground faults. The ground fault current will be limited in these systems to a low level. This will protect the cabling from serious damage from ground faults.

Now that we have our system specifications, we will now run our calculations to design it. Our design is the same as the commercial design from the previous chapter and we will now expand that design to utility scale.

We will combine these ten inverters using an electrical switching center. This will give:

Switching center output current = 10 x inverter current

= 10 x 301 amps

= 3,010 amps

Our minimum fuse size is 125% larger than the circuit current:

Cable amperage = 125% x 3010A

= 3,762.5A

NEC240.6(A) indicates a standard fuse size of 4,000A for this circuit.

This seems like a lot of current, but we are traveling just a short distance to the transformer that is mounted on the same equipment pad, so our cable length is minimal. Let's see how this compares with our de-rated values:

De-rate the current for an ambient temperature of 40°C:

De-rated current = 4,000A x 0.91

= 4,396A

Looking at NEC table 310.16 we see that 90°C 2000 kcmil copper cable can carry 750 amps. So we will need to parallel up the cables to achieve the desired current:

Number of parallel cables needed = de-rated current / 750 amps

= 4,396A / 750A

= 5.87

We round up to 6 parallel cables of 90°C 2000 kcmil.

Cable current = Switching center output current /6

= 3010A / 6

= 502A per cable

We feed into the transformer 480 volt winding. On the output we get 24,000 volts. This gives:

Transformer step up = 24,000V / 480V

= 50

Since we have increased the voltage by 50, we have correspondingly reduced the current by 50. We also need to allow for the efficiency of the transformer at 99%.

MV current = Switching center current x 99% / 50

= 59.598 Amps

This gives an MV power of:

MV Transformer Power = sqrt(3) x 24,000V x 59.598A

= 2,477,443W

We come out of the transformer to a MV fused switch. We apply 125% to the current to get:

MV fuse size = 125% x MV fuse current

= 125% x 59.598A

= 75A

NEC240.6(A) indicates a standard fuse size of 80A.

We would use the larger de-rating of either 125% or the de-rating for a 60°C temperature if the fuse manufacturer requires it.

80A is our fuse size and also our minimum cable size. Our cable will be in an underground duct, so we will look into using the following cable per NEC table 310.77:

MV Cable = 90°C MV90 6AWG copper cable = 90A

This is rated at 20C ground ambient temperature. We are using 40°C for the ambient ground temperature, so lets check on that for de-rating using formula NEC 310.60-C-4:

Ambient current= cable rated current x sqrt((Conductor temperature x desired cable ambient temperature x dielectric loss temperature rise) / (Conductor temperature x cable ambient temperature from table x dielectric loss temperature rise)

= 90 x sqrt ((90 - 40 – 20) / (90 - 20 – 20))

= 69A

As we can see, this cable is not large enough for our purpose after de-rating, so we move onto the next cable size. This is 4AWG 90°C type MV 90 copper at 115A.

= 115 x sqrt ((115 - 40 – 20) / (115 - 20 – 20))

= 98A

So 4AWG is our choice for the 24kV 2.5MW cabling.

Now onto the 10MW 24kV cabling.

10 MW current = 10MW / sqrt(3) x 24,000V

= 240.57 Amps

To obtain our cable size, we apply 125% to the current to get:

10MW fuse size = 125% x MV current

= 125% x 240.57A

= 301A

NEC table 310.77 indicates that 250 kcmil 90°C MV90 copper cable can carry 325A. Now we de-rate this for an underground temperature of 40°C to obtain our cable size.

= 325A x sqrt ((325A - 40 – 20) / (325A - 20 – 20))

= 313.3A

This is above the fuse rating, and we will use this cable.

Our switchgear in the AC circuit will be selected based on the interrupt current supplied by the utility. The equipment grounding will be done in accordance with the equipment manufacturers recommendations and local codes.

So here are our system specifications:

- 10 MVA Cabling: 90°C 250 kcmil type MV-90 copper at 24 kV

- 1 of: 10 MVA, 24 kVAC distribution center with ground fault detection

- 2.5MVA cabling: 90°C 4AWG type MV-90 copper at 24 kV

- 4 of: 2.5MVA, 24 kVAC medium voltage pad mounted fused switches with 80A fuses

- 4 of: medium voltage transformers: Dry type, 99% efficiency, 2.5MVA, 24 kVAC / 480VAC

- 4 of: 2.5 MVA 480VAC switching centers with ground fault detection and ten 400A circuit breakers

- 480VAC switching center cables 90°C 2000 kcmil copper

- 40 of: Inverter 250 kW, 3 phase, 480 VAC, 600 VDC

- Inverter 480VAC cabling: 90°C 500 kcmil copper

- Inverter DC cabling for 10 input combiner: 90°C 1/0 AWG copper

- Inverter equipment ground cabling for 10 input combiner: 90°C 2 AWG copper

- 160 of: 10 input combiner boxes

- Inverter DC cabling for 9 input combiner: 90°C 1 AWG copper

- Inverter equipment ground cabling for 9 input combiner: 90°C 3 AWG copper

- 160 of: 9 input combiner boxes

- Solar string cabling: 90°C 14 AWG copper

- 3,040 of: 15A DC solar photovoltaic string fuses

- 42,560 of: solar modules

- Row Spacing: 26.7 feet

When the utility provided the interconnection details you will be able to calculate the fault currents in the design.

With these values the correct switchgear can be selected, ensuring that it is rated for back feed.

Once the design is completed, it will need to be stamped by a licensed Professional Engineer to verify the design and the calculations.

The following pages show our design:

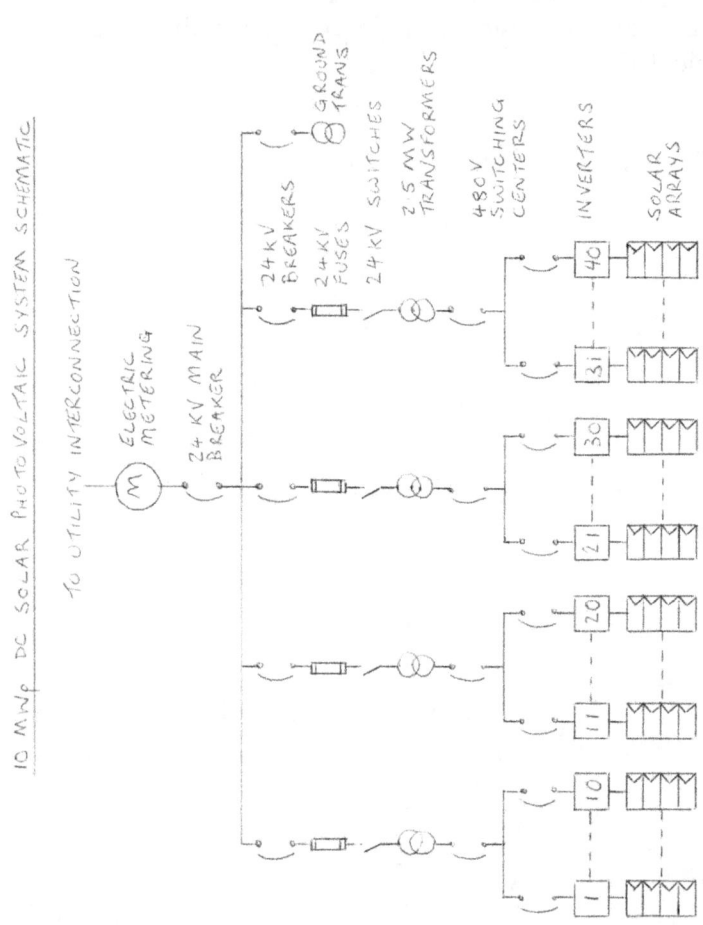

10 MWp DC SOLAR PHOTOVOLTAIC SYSTEM SCHEMATIC

10 MVA DISTRIBUTION

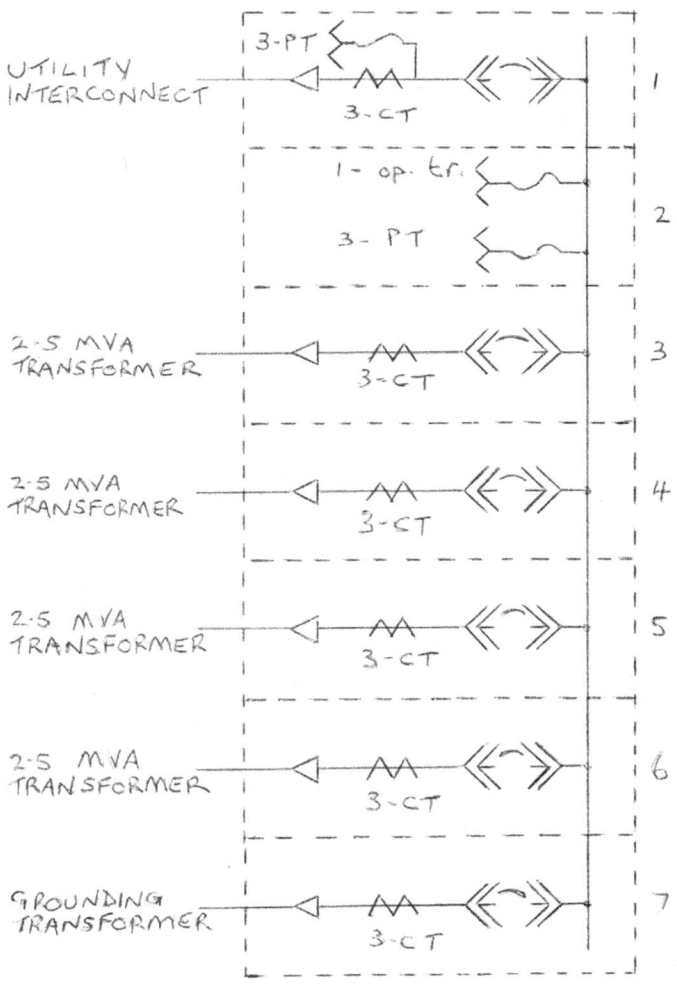

Solar Photovoltaic Design by Steven Magee

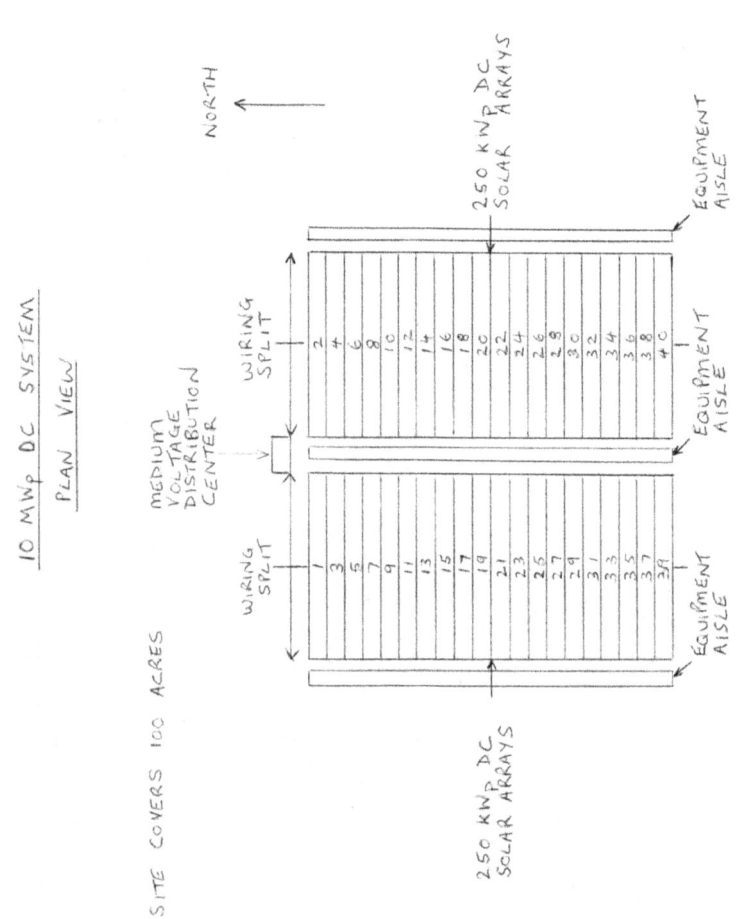

88

Solar Photovoltaic Design by Steven Magee

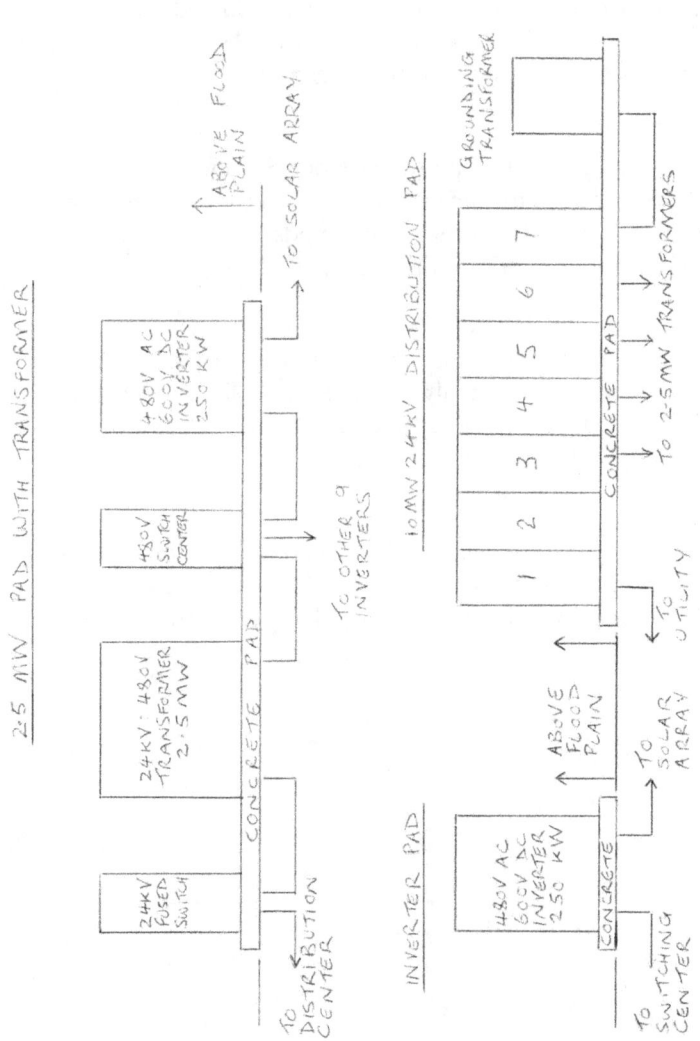

89

9. Summary

Solar photovoltaics is a rapidly changing field and new technology is constantly being developed. As such, always design from the latest electrical and building codes.

In the USA, the National Electric Code Section 690 Solar Photovoltaic Systems details the latest electrical design standards. The USA National Electric Code is updated every three years and it is important to design from the latest version.

Follow the design and installation manuals for the products used in your design. As ever, always have any designs independently verified by a registered Professional Engineer that specializes in solar photovoltaics.

I hope that the book has given you a good overview of solar photovoltaic design for grid interconnected systems and helps with designing your systems to your local solar photovoltaic codes.

References

- NFPA National Electrical Code 2005
- NFPA National Electrical Code 2008
- IEEE/ANSI-C2-2007 National Electric Safety Code
- Occupational Safety and Health Administration www.osha.gov
- National Renewable Energy Laboratories www.nrel.gov
- United States Department of Energy www.energy.gov

Author Contact

Steven Magee,
3618 S. Desert Lantern Road,
Tucson,
AZ 85735
USA

I hope that you found the book informative and please let me know about any questions or comments about the book.

I am a consultant on new solar photovoltaic projects, solar photovoltaic troubleshooting and solar photovoltaic training. Please feel free to contact me for any help or assistance in these areas.

www.ingramcontent.com/pod-product-compliance
Lightning Source LLC
Chambersburg PA
CBHW071242170526
45165CB00003B/1202